~A BINGO BOOK~

Life Science Bingo Book

COMPLETE BINGO GAME IN A BOOK

Written By Rebecca Stark

Educational Books 'n' Bingo

Educational Books 'n' Bingo

ISBN 978-0-87386-544-9

Printed in the U.S.A.

LIFE SCIENCE BINGO
Directions

INCLUDED:

List of Terms

Templates for Additional Terms and Clues

2 Clues per Term

30 Unique Bingo Cards

Markers

1. **Either cut apart the book or make copies of ALL the sheets. You might want to make an extra copy of the clue sheets to use for introduction and review. Keep the sheets in an envelope for easy reuse.**

2. Cut apart the call cards with terms and clues.

3. Pass out one bingo card per student. There are enough for a class of 30.

4. Pass out markers. You may cut apart the markers included in this book or use any other small items of your choice.

5. Decide whether or not you will require the entire card to be filled. Requiring the entire card to be filled provides a better review. However, if you have a short time to fill, you may prefer to have them do the just the border or some other format. Tell the class before you begin what is required.

6. There are 50 topics. Read the list before you begin. If there are any topics that have not been covered in class, you may want to read to the students the topic and clues before you begin.

7. There is a blank space in the middle of each card. You can instruct the students to use it as a free space or you can write in answers to cover topics not included. Of course, in this case you would create your own clues. (Templates provided.)

8. Shuffle the cards and place them in a pile. Two or three clues are provided for each topic. If you plan to play the game with the same group more than once, you might want to choose a different clue for each game. If not, you may choose to use more than one clue.

9. Be sure to keep the cards you have used for the present game in a separate pile. When a student calls, "Bingo," he or she will have to verify that the correct answers are on his or her card AND that the markers were placed in response to the proper questions. Pull out the cards that are on the student's card keeping them in the order they were used in the game. Read each clue as it was given and ask the student to identify the correct answer from his or her card.

10. If the student has the correct answers on the card AND has shown that they were marked in response to the *correct questions,* then that student is the winner and the game is over. If the student does not have the correct answers on the card OR he or she marked the answers in response to *the wrong questions,* then the game continues until there is a proper winner.

11. If you want to play again, reshuffle the cards and begin again.

Have Fun!

TERMS INCLUDED

ADAPTATIONS

ALGAE

ANATOMY

ANIMAL

BACTERIA

BIOLOGY

BIOTIC FACTORS

BOTANY

BRAIN

CARBON

CARNIVORE

CELL

CIRCULATORY SYSTEM

DIGESTIVE SYSTEM

DNA

EAR

ENDANGERED SPECIES

ENDOCRINE SYSTEM

EVOLUTION

EXTINCT

FOOD CHAIN

FUNGI

GENE   X

GENETICS

HEART

HERBIVORE

INVERTEBRATES

LIFE CYCLE

CAROLUS LINNAEUS

MAMMALS

MICROBIOLOGY

MICROSCOPE

MUSCLES

NERVOUS SYSTEM

ORGANS

PHOTOSYNTHESIS

PLANTS

PRIMATES

PROTOZOA

REPRODUCTION

RESPIRATORY SYSTEM

SENSES

SKELETAL SYSTEM

SKIN

SYSTEMS

TISSUE

URINARY SYSTEM

VERTEBRATES

VIRUSES

ZOOLOGY

Additional Terms

Choose as many additional terms as you would like and write them in the squares.
Repeat each as desired.
Cut out the squares and randomly distribute them to the class.
Instruct the students to place their square on the center space of their card.

Life Science Bingo

Clues for
Additional Terms

Write three clues for each of your additional terms.

_____	_____
1.	1.
2.	2.
3.	3.
_____	_____
1.	1.
2.	2.
3.	3.
_____	_____
1.	1.
2.	2.
3.	3.

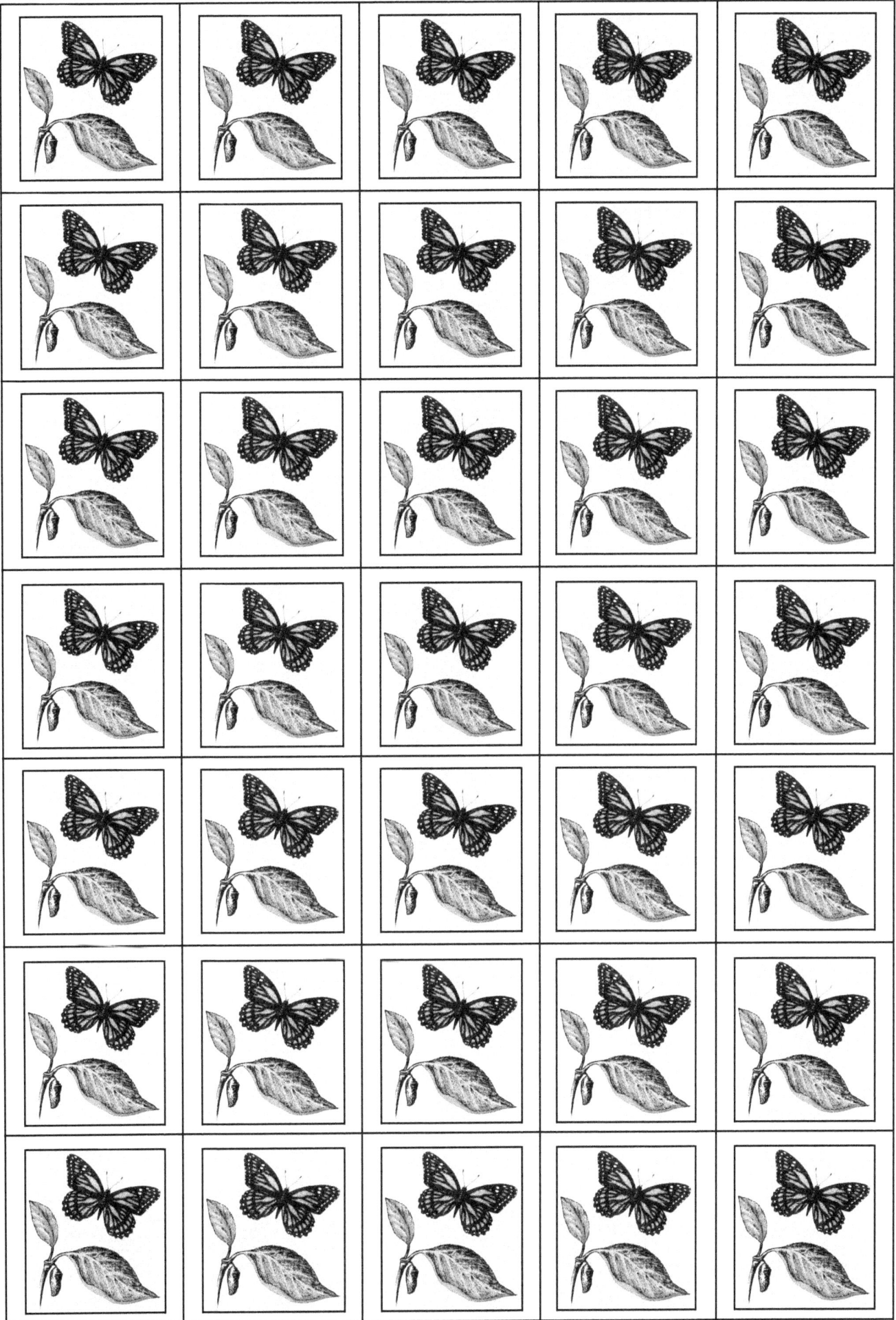

ADAPTATIONS
1. These features in the structure or function of an organism allow it to survive and reproduce more effectively in its environment.
2. Sometimes they are due to an effort to cope with changing seasons.
3. Some occur as a result of an animal evolving into a more efficient predator.

ALGAE
1. The largest, most complex forms are called seaweeds.
2. These plant-like, photosynthetic organisms grow in water.
3. Some of these plant-like organisms are one-celled and some are multi-celled.

ANATOMY
1. It is the study of animal form, especially the human body.
2. It is the science of the shape and structure of organisms.
3. This subdivision of morphology deals with the structure of animals. (Morphology is the branch of biology that deals with the form and structure of organisms without considering function.)

ANIMAL
1. It is a living organism that has no cell walls. Most have voluntary movement.
2. Members of this kingdom include invertebrates and vertebrates. Vertebrates comprise fish, birds, amphibians, reptiles and mammals.
3. Members of this kingdom are heterotrophic; they obtain nourishment from the ingestion and breakdown of organic matter.

BACTERIA
1. These single-celled organisms have a cell membrane and cytoplasm but no cell nucleus. They multiply by simple division.
2. Unlike other life forms, their genetic material, or DNA, is not enclosed in a cellular compartment called the nucleus.
3. Some, like *E. coli*, cause disease. Others, like those that help us digest our food, are helpful.

BIOLOGY
1. This science is concerned with the structure, function, distribution, adaptation, interactions, and evolution of plants and animals.
2. Botany is the branch that studies plant life. Zoology is the branch that studies animal life.
3. Genetics and anatomy are two branches of this science.

BIOTIC FACTORS
1. They are the living parts of the ecosystem with which an organism must interact.
2. These include everything that results from the activities of living things that directly or indirectly affect an organism in its environment.
3. Examples would be an organism's prey and predators.

BOTANY
1. This subdivision of biology is the science of plants.
2. One who specializes in this science would be called a botanist.
3. This includes the study of plants, algae, and fungi.

BRAIN
1. This organ controls the human nervous system.
2. Its largest part, the cerebrum, is where most of our important mental functions take place.
3. Its cerebellum is involved with the regulation and coordination of voluntary movement, posture, and balance.

CARBON
1. There are almost ten million known ____ compounds, thousands of which are vital to organic and life processes.
2. Without this element, life as we know it would be impossible.
3. No life has been observed that is not based upon this element.

Life Science Bingo

© Barbara M. Peller

CARNIVORE
1. It is a predatory, flesh-eating animal.
2. It eats mostly herbivores.
3. Examples of this kind of animal that eats other animals include eagles, tigers, alligators and most humans.

CELL
1. It is the basic structural and functional unit in all living things.
2. It is the smallest structural unit of living matter able to function independently.
3. It is the basic microscopic unit of living things.

CIRCULATORY SYSTEM
1. It comprises the heart, blood and blood vessels.
2. This system is responsible for moving blood throughout the body, thereby transporting oxygen and nutrients to the body's cells and disposing waste material.
3. The three types of blood vessels in this system are arteries, veins and capillaries.

DIGESTIVE SYSTEM
1. This system is responsible for breaking down food into nutrients.
2. Some important organs in this system are the mouth, pharynx, esophagus, stomach, small intestine, and large intestine.
3. The main organs in this system make up the alimentary canal.

DNA
1. It is an acronym for deoxyribonucleic acid.
2. It is the genetic material in humans and most other organisms.
3. It is a double-stranded molecule held together by weak hydrogen bonds between base pairs of subunits, called nucleotides. The molecule forms a double helix in which two strands spiral about one other.

EAR
1. This organ is responsible for audition. It perceives sound by detecting vibrations.
2. The inner part of this organ is helpful in maintaining balance.
3. There are three tiny bones in its tympanic cavity: the malleus, or hammer; the incus, or anvil; and the stapes, or stirrup.

ENDANGERED SPECIES
1. It is the population of an organism at risk of becoming extinct.
2. The American peregrine falcon was on this list from 1970 to 1999. It was taken off because it recovered. The Mariana mallard was on the list from 1977 to 2004, when it became extinct.
3. The right whale, the ocelot and the leatherback sea turtle are examples.

ENDOCRINE SYSTEM
1. This system helps maintain a stable internal environment by sending out chemical signals in the form of hormones.
2. Glands in this system are called ductless; hormones go directly into the bloodstream.
3. Among the glands in this system are the pituitary gland, the thyroid gland, the parathyroid gland and the adrenal gland.

EVOLUTION
1. It is the process of change by which new species develop from preexisting species over time.
2. Charles Darwin, a 19th-century naturalist, is considered the "Father of ___."
3. Darwin defined it as "descent with modification."

EXTINCT
1. This term describes a species or a population that no longer exists.
2. Dinosaurs became this about 65 million years ago.
3. The dodo bird and the Tasmanian wolf are examples of animals who are now this.

Life Science Bingo

© Barbara M. Peller

FOOD CHAIN

1. This describes the feeding relationships between species within an ecosystem.
2. Plankton, microscopic plant life that floats in the ocean, is the first level of this.
3. It usually starts with a primary producer and ends with a carnivore.

FUNGI

1. These organisms obtain food by breaking down dead organic material and produce spores.
2. These organisms lack chlorophyll and are usually non-mobile. At one time they were classified as plants.
3. Some examples are molds, mildews, yeasts and mushrooms.

GENE

1. It is the basic biological unit of heredity.
2. It is a segment of DNA at a specific location on a chromosome.
3. This hereditary unit determines a particular characteristic in an organism.

GENETICS

1. This is the science of heredity.
2. This branch of biology focuses on how particular traits are transmitted from parents to children.
3. A professional who specializes in this field of study is called a geneticist.

HEART

1. This important organ pumps blood throughout the body. It is part of the circulatory system.
2. This organ is divided into 2 sections separated by the septum. The upper chamber of each section is called an atrium; the lower is called a ventricle.
3. Its right ventricle pumps blood to the lungs. Its left pumps blood to the rest of the body.

HERBIVORE

1. This kind of animal feeds on plants.
2. It is the opposite of "carnivore."
3. This kind of animal needs more energy than a carnivore. Many, like cows and sheep, eat all day long.

INVERTEBRATES

1. These animals lack a vertebral column, or backbone.
2. Insects, spiders, shrimp and jellyfish are among these animals without an internal skeleton made of bone.
2. This group without vertebrae comprise about 98% of all animal species.

LIFE CYCLE

1. This refers to the successive stages through which an organism passes.
2. Some insects, like butterflies, go through complete metamorphosis, or change, during theirs. The four stages in a butterfly's ___ are egg, larva, pupa and adult.
3. The aquatic larval stage in the ___ of an amphibian is called a tadpole.

CAROLUS LINNAEUS (Carl von Linné)

1. This Swedish botanist is sometimes called the "Father of Taxonomy."
2. He established binomial nomenclature, a scientific way of naming plants and animals.
3. In his method of naming plants and animals, each scientific name has 2 Latin parts: the genus, and the species. The scientific name for humans, for example, is *Homo sapiens*.

Life Science Bingo

MAMMALS

1. These warm-blooded vertebrates have hair. The females produce milk with which to nourish their young.
2. Monkeys, rabbits, bats, elephants, whales, and kangaroos are all examples.
3. Woolly mammoths and saber-toothed tigers are extinct examples.

MICROBIOLOGY

1. This is the branch of science that studies organisms too small to be seen with the naked eye.
2. This branch of life science includes the study of protozoans, algae, bacteria, and viruses.
3. Anton van Leeuwenhoek, who is known for his improvement of the microscope, is often called the father of this branch of science.

MICROSCOPE

1. This optical instrument utilizes one or more lenses to magnify images of things too small to be seen by the naked eye.
2. An electron ___ uses electrons rather than visible light to produce magnified images.
3. Anton van Leeuwenhoek, considered the first microbiologist, is known for his improvement of this equipment.

MUSCLES

1. There are three kinds: skeletal, smooth and cardiac.
2. The only voluntary ones are the skeletal ones, which are attached to our bones.
3. They are specialized to contract, or shorten. It is this contracting that allows them to do their job and move the various body parts.

NERVOUS SYSTEM

1. The central ___ comprises the brain and the spinal cord.
2. The peripheral ___ comprises all the nerves that are spread throughout the body.
3. The somatic ___ is associated with voluntary control of body movements; the autonomic ___ maintains normal internal functions and is not subject to voluntary control.

ORGANS

1. These are groups of tissues working together to perform a specific function or group of functions.
2. Some human ___ are the heart, the lungs, the bones, the skin and the stomach.
3. Plant ___ include the root, the leaf, the stem, the seed, the flower and the fruit.

PHOTOSYNTHESIS

1. This is the process by which green plants use water and carbon dioxide to create glucose and oxygen.
2. This process takes energy from the sun and converts it into a storable form which plants use for their own life processes.
3. Animals provide the carbon dioxide needed for this process and get oxygen in return.

PLANTS

1. These include trees, herbs, bushes, grasses, vines, ferns, mosses, and green algae.
2. They are the major producers in an ecosystem.
3. Unlike animals, they lack organs for mobility.

PRIMATES

1. Most members of this order of mammals have binocular vision, refinement of hands and feet for grasping, and enlarged cerebral hemispheres.
2. Humans, apes, and monkeys belong to this order.
3. Lemurs and other prosimians are a more primitive suborder of this order.

PROTOZOA

1. These single-celled animals are larger and more complex than bacteria. Their cell contains a distinct membrane-bound nucleus.
2. Like most algae, they are considered protists. In other words, they are not considered true animals, plants or fungi.
3. Amoebas and ciliates are ___.

REPRODUCTION

1. It is the biological process by which new individual organisms are produced.
2. This process is grouped into two main types: sexual and asexual.
3. Most plants have the ability to achieve this asexually.

Life Science Bingo

RESPIRATORY SYSTEM 1. The main function of this system is to supply the blood with oxygen. 2. This system gets oxygen to the blood through an exchange of gases. When we breathe, we inhale oxygen and exhale carbon dioxide. 3. Organs of this system include the mouth, the nose, the trachea, the lungs & the diaphragm.	**SENSES** 1. The five include sight, hearing, taste, touch, and smell. 2. Each of the five specializes in a particular type of information. This information is sent to the brain by the way of nerve cells. 3. The ___ of touch originates in the bottom layer of skin, called the dermis.
SKELETAL SYSTEM 1. This system is made up of bones and connective tissues, such as cartilage, ligaments, and tendons. It is sometimes grouped together with the muscular system. 2. This system provides support, allows movement and protects our vital organs. 3. The skull, which is part of this system, protects our brain.	**SKIN** 1. The branch of medicine dealing with this organ is called dermatology. 2. This large organ acts as a barrier against infection and provides a sense of touch. 3. This organ regulates body temperature and excretes waste product and excess salt from the body.
SYSTEMS 1. Group of organs working together form these. 2. The digestive, muscular, skeleton, nervous, and circulatory are some in the human body. 3. Cells work together to form tissues; tissues work together to form organs; and organs work together to form these.	**TISSUE** 1. It is a group of cells that have a similar structure and which function together as a unit. 2. There are four types: connective, epithelial, muscle and nervous. 3. When different kinds work together, they make an organ.
URINARY SYSTEM 1. It is also called the excretory system. 2. Its main organs are the kidneys and bladder. 3. The organs of this system control how much water and salts are absorbed back into the bloodstream and what is taken out as waste. It also filters the blood.	**VERTEBRATES** 1. These are animals with a backbone, or spinal column. 2. These include fishes, amphibians, reptiles, birds, and mammals. 3. Their spinal column forms part of a complete internal skeleton. All members of this subphylum possess a cranium.
VIRUSES 1. These sub-microscopic infectious agents do not have cells and are unable to grow or reproduce outside a host cell. 2. Common diseases caused by them include the common cold, influenza, and chickenpox. 3. Biologists debate whether or not these sub-microscopic infectious agents are living organisms. Life Science Bingo	**ZOOLOGY** 1. It is the branch of biology that deals with animals and animal life. 2. Scientists who specialize in this field are called zoologists. 3. This branch of science focuses on the structure, physiology, development, and classification of animals.

Life Science Bingo

Evolution	Adaptations	Biology	Heart	Carolus Linnaeus
Algae	Anatomy	Systems	Microbiology	Viruses
Animal	Urinary System		Microscope	Circulatory System
Tissue	Endocrine System	Zoology	Invertebrates	Nervous System
Plants	Carbon	Reproduction	Vertebrates	Skeletal System

Life Science Bingo

Tissue	Algae	Mammals	Respiratory System	Genetics
Nervous System	Endangered Species	Animal	Cell	Herbivore
Digestive System	Carbon		Extinct	Zoology
Protozoa	Organs	Urinary System	Photosynthesis	Skeletal System
Viruses	Systems	Reproduction	Carnivore	Vertebrates

Life Science Bingo

Tissue	Zoology	Endangered Species	Invertebrates	Algae
Microbiology	Anatomy	Botany	Adaptations	Cell
Endocrine System	Systems		Herbivore	Bacteria
Urinary System	Digestive System	Plants	Protozoa	Mammals
Vertebrates	Carnivore	Reproduction	Photosynthesis	Genetics

Life Science Bingo

Urinary System	Herbivore	Biology	Cell	Genetics
Life Cycle	Biotic Factors	Adaptations	Respiratory System	Algae
Microscope	Protozoa		Carolus Linnaeus	Heart
Zoology	DNA	Systems	Reproduction	Animal
Carnivore	Viruses	Organs	Vertebrates	Circulatory System

Life Science Bingo: Card No. 4

Life Science Bingo

Viruses	Carolus Linnaeus	Endocrine System	Animal	Carnivore
Life Cycle	Zoology	Botany	Extinct	Anatomy
Biology	Circulatory System		Microbiology	Genetics
Skeletal System	Gene	Evolution	Photosynthesis	Ear
Endangered Species	Reproduction	Algae	Urinary System	Microscope

Life Science Bingo

Bacteria	Herbivore	Mammals	Genetics	Circulatory System
Invertebrates	Endocrine System	Ear	Adaptations	Algae
Respiratory System	Cell		Biotic Factors	Extinct
Reproduction	Plants	Photosynthesis	Organs	Biology
Nervous System	Animal	Evolution	Microscope	DNA

Life Science Bingo

Evolution	Herbivore	Gene	Microbiology	Endangered Species
Nervous System	Genetics	Carbon	Anatomy	Life Cycle
Mammals	Heart		Extinct	Biotic Factors
Urinary System	Protozoa	Botany	Tissue	Digestive System
Reproduction	Carnivore	Photosynthesis	Organs	Bacteria

Life Science Bingo

Microscope	Herbivore	Fungi	Invertebrates	Biotic Factors
Life Cycle	Biology	Respiratory System	Circulatory System	Animal
DNA	Primates		Genetics	Carolus Linnaeus
Vertebrates	Urinary System	Tissue	Cell	Protozoa
Systems	Reproduction	Organs	Endocrine System	Nervous System

Life Science Bingo

Extinct	Endangered Species	Carbon	DNA	Carnivore
Cell	Genetics	Microscope	Endocrine System	Herbivore
Brain	Evolution		Anatomy	Fungi
Ear	Skeletal System	Plants	Microbiology	Gene
Protozoa	Photosynthesis	Botany	Tissue	Carolus Linnaeus

Life Science Bingo

Tissue	Invertebrates	Biotic Factors	Respiratory System	DNA
Circulatory System	Animal	Adaptations	Anatomy	Genetics
Primates	Herbivore		Heart	Digestive System
Plants	Skeletal System	Ear	Photosynthesis	Brain
Botany	Nervous System	Mammals	Viruses	Microscope

Life Science Bingo: Card No. 10

Life Science Bingo

Bacteria	Herbivore	Endocrine System	Ear	Nervous System
Fungi	Brain	Microbiology	Extinct	Adaptations
Life Cycle	Genetics		Mammals	Carbon
Botany	States' Rights	Photosynthesis	Carnivore	Tissue
Cell	Reproduction	Evolution	Organs	Endangered Species

Life Science Bingo: Card No. 11

Life Science Bingo

Endangered Species	Carolus Linnaeus	Brain	Invertebrates	Extinct
Carbon	Nervous System	Biology	Organs	Anatomy
Evolution	Gene		Circulatory System	Respiratory System
Reproduction	Protozoa	Genetics	Tissue	Life Cycle
Herbivore	Fungi	Primates	Cell	Animal

Life Science Bingo

Ear	Carolus Linnaeus	Bacteria	Brain	Circulatory System
Biology	Fungi	Gene	Extinct	Digestive System
Invertebrates	Endangered Species		Carbon	Genetics
Microscope	Photosynthesis	Biotic Factors	Primates	Tissue
Reproduction	Skeletal System	Organs	Evolution	Microbiology

Life Science Bingo: Card No. 13

Life Science Bingo

Carnivore	Gene	Endocrine System	Extinct	Cell
Animal	Evolution	Brain	Anatomy	Herbivore
Ear	Heart		Mammals	Botany
Skeletal System	Photosynthesis	Primates	Biotic Factors	Bacteria
Reproduction	Respiratory System	Digestive System	Nervous System	Microscope

Life Science Bingo: Card No. 14

© Barbara M. Peller

Life Science Bingo

Microbiology	Extinct	Endocrine System	Endangered Species	Invertebrates
Bacteria	Mammals	Adaptations	Biology	Cell
Circulatory System	Evolution		Algae	Herbivore
Reproduction	Brain	Fungi	Photosynthesis	Ear
Nervous System	Protozoa	Organs	DNA	Carbon

Life Science Bingo

Biotic Factors	Brain	Fungi	DNA	Senses
Respiratory System	Digestive System	Gene	Life Cycle	Heart
Ear	Carolus Linnaeus		Circulatory System	Carbon
Urinary System	Animal	Reproduction	Muscles	Tissue
Cell	Skin	Organs	Protozoa	Herbivore

Life Science Bingo

Botany	Muscles	Food Chain	Brain	Carnivore
Microbiology	Cell	Photosynthesis	Heart	Gene
Extinct	Microscope		Skin	Fungi
Skeletal System	Nervous System	Tissue	Endocrine System	Digestive System
Plants	Ear	Endangered Species	Invertebrates	Carolus Linnaeus

Life Science Bingo

DNA	Primates	Animal	Ear	Respiratory System
Herbivore	Botany	Plants	Circulatory System	Cell
Extinct	Digestive System		Food Chain	Biology
Skeletal System	Adaptations	Photosynthesis	Tissue	Mammals
Skin	Brain	Endocrine System	Muscles	Bacteria

Life Science Bingo

Circulatory System	Bacteria	Brain	Fungi	Circulatory System
Microbiology	Invertebrates	Herbivore	Endangered Species	Heart
Muscles	Carnivore		Anatomy	Algae
Mammals	Skin	Plants	Protozoa	Food Chain
Biology	Senses	Nervous System	Microscope	Organs

Life Science Bingo: Card No. 19

Life Science Bingo

Primates	Muscles	Invertebrates	Brain	Organs
Animal	Carbon	Life Cycle	Plants	Respiratory System
Carolus Linnaeus	Gene		Urinary System	Adaptations
Viruses	Systems	Vertebrates	Protozoa	Skin
Zoology	Microscope	Senses	Tissue	Food Chain

Life Science Bingo

Microbiology	Bacteria	Life Cycle	Brain	Viruses
Carolus Linnaeus	Food Chain	Biotic Factors	Fungi	Evolution
Digestive System	Nervous System		Muscles	Endocrine System
Plants	Endangered Species	Skin	Skeletal System	Microscope
Urinary System	Senses	Organs	Botany	Protozoa

Life Science Bingo

DNA	Mammals	Food Chain	Biology	Ear
Respiratory System	Invertebrates	Algae	Fungi	Anatomy
Animal	Heart		Evolution	Gene
Skin	Skeletal System	Protozoa	Adaptations	Carnivore
Senses	Botany	Muscles	Digestive System	Life Cycle

Life Science Bingo: Card No. 22

© Barbara M. Peller

Life Science Bingo

Biotic Factors	Muscles	Endangered Species	Biology	Organs
Bacteria	Primates	Nervous System	Microbiology	Adaptations
Mammals	Ear		Vertebrates	Evolution
Digestive System	Senses	Skin	Botany	Protozoa
Viruses	Systems	Microscope	Plants	Food Chain

Life Science Bingo

Biotic Factors	Primates	Carnivore	Muscles	Fungi
Food Chain	Organs	Life Cycle	Respiratory System	Evolution
Gene	DNA		Ear	Digestive System
Viruses	Vertebrates	Skin	Botany	Carolus Linnaeus
Zoology	Urinary System	Senses	Invertebrates	Systems

Life Science Bingo

Urinary System	Life Cycle	Muscles	Endocrine System	Food Chain
Adaptations	Skeletal System	Microbiology	Biotic Factors	Anatomy
Carolus Linnaeus	Fungi		Vertebrates	Skin
Algae	Viruses	Systems	Senses	Heart
Organs	Carnivore	Animal	Cell	Zoology

Life Science Bingo: Card No. 25

Life Science Bingo

Food Chain	Muscles	Mammals	Respiratory System	DNA
Plants	Invertebrates	Fungi	Primates	Biotic Factors
Skeletal System	Vertebrates		Heart	Urinary System
Botany	Biology	Viruses	Senses	Skin
Gene	Cell	Endocrine System	Systems	Zoology

Life Science Bingo

Mammals	Animal	Muscles	Primates	Carbon
Viruses	Vertebrates	Microbiology	Skin	Anatomy
Photosynthesis	Systems		Senses	Urinary System
DNA	Bacteria	Life Cycle	Zoology	Adaptations
Cell	Heart	Food Chain	Algae	Gene

Life Science Bingo

Circulatory System	Primates	Algae	Muscles	Biotic Factors
Carbon	Food Chain	Vertebrates	Respiratory System	Heart
Systems	Digestive System		Gene	Plants
Tissue	DNA	Nervous System	Senses	Skin
Biology	Extinct	Cell	Zoology	Viruses

Life Science Bingo

Food Chain	Primates	DNA	Microbiology	Extinct
Skeletal System	Plants	Life Cycle	Genetics	Algae
Carolus Linnaeus	Vertebrates		Anatomy	Muscles
Endocrine System	Viruses	Gene	Senses	Skin
Cell	Fungi	Zoology	Bacteria	Systems

Life Science Bingo: Card No. 29

Life Science Bingo

Carnivore	Muscles	Respiratory System	Extinct	Skin
Adaptations	Primates	Mammals	Heart	Anatomy
Skeletal System	Ear		Genetics	Carbon
Zoology	Bacteria	Biology	Senses	Vertebrates
Viruses	Endangered Species	Systems	Food Chain	Algae